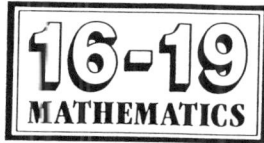

Modelling with differential equations

Unit guide

The School Mathematics Project

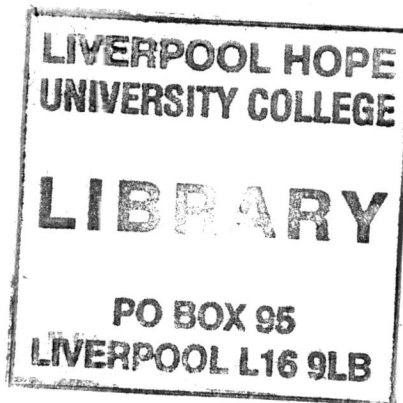

CAMBRIDGE
UNIVERSITY PRESS

Main authors	Ann Kitchen
	Kevin Lord
	Mike Savage
	Nigel Webb
	Julian Williams
Team leaders	Kevin Lord and Julian Williams
Project director	Stan Dolan

The authors would like to give special thanks to Ann White for her help in preparing this book for publication.

Published by the Press Syndicate of the University of Cambridge
The Pitt Building, Trumpington Street, Cambridge CB2 1RP
40 West 20th Street, New York, NY 10011-4211, USA
10 Stamford Road, Oakleigh, Victoria 3166, Australia

First published 1993

Produced by 16-19 Mathematics and Laserwords, Southampton

Printed in Great Britain by Scotprint Ltd., Musselburgh.

2087/11

ISBN 0 521 42653 7

Contents

Introduction to the unit
(for the teacher)

Prior knowledge of *Modelling with circular motion* or *Modelling with rigid bodies* is not assumed, but students must already have covered *Newton's laws of motion* and *Modelling with force and motion*.

It is recognised that many students working through this unit may be doing so without the benefit of substantial contact time with a teacher. The unit has therefore been written to facilitate 'supported self-study'. It is assumed that even a minimal allocation of teacher time will allow contact at the start and end of each chapter and so:

- solutions to all thinking points and exercises are in the students' text;

- a substantial discussion point in one of the opening sections enables the teacher to introduce each chapter;

- a special tutorial sheet can be used to focus discussion at a final tutorial on the work of the chapter.

It is assumed that students will already have studied the unit *Differential equations* or that they will be studying it at the same time. The algebraic and numerical techniques of solving the differential equations in this unit are all introduced there.

This unit will develop the skills of modelling dynamic situations using Newtonian mechanics. The students will learn to set up a model, derive suitable differential equations and interpret and validate the solutions appropriately in their real-life contexts. This will involve application of the techniques of solving differential equations. The students are expected to use analytical methods where appropriate for solving first and second order equations, i.e. separation of variables and integration for first order equations, and the use of a complementary function and particular integral for second order equations. For particular initial conditions, and where analytical methods fail, the students are expected to use an Euler step method for numerical integration of first or second order equations.

The use of a graphic calculator is essential for sketching graphs of solutions and for applying numerical methods of integration. Suitable programs are given at the end of the student text.

The use of a computer is essential for sketching 'direction field' diagrams. The *16-19 Mathematics* program *Solution sketcher* was specially written to help with this. A computer may also be used to implement programs to calculate numerical solutions using Euler step methods, either using a programming language or a spreadsheet.

Practical work is involved in finding models for resistive drag forces and for tensions in springs and strings. These may already have been covered by the students in *Modelling with force and motion*. Throughout the unit there will be opportunities to validate results practically using experiments, practical simulations or through research data. These efforts should be encouraged. There are many suggestions for modelling investigations throughout the text, exercises and thinking points and it is expected that practical work will play an important part in these.

In most modelling investigations, an initial model can be set up that will lead to a simple differential equation whose solution will involve standard analytical methods. However, further extension and refinement of the model may be necessary and might lead to more difficult or even 'insoluble' (by traditional analytical methods) differential equations. Students are then expected to apply numerical methods which will involve the use of programs on calculators or computers.

Chapters 1 and 2

These chapters look at applications of resisted motion in one dimension. Models for air resistance are investigated. Newton's second law is used to derive first order differential equations of motion. These are solved by separation of variables and by numerical methods. Graphs of solutions help to develop an understanding of terminal speed.

Chapter 3

This chapter introduces models of motion for a particle under variable gravity and for a rocket of variable mass. These also lead to first order differential equations.

Chapter 4

The standard model of simple harmonic motion for mass-spring systems is covered. This involves the solution of the simple harmonic motion equation, which may be quoted and matched to particular initial conditions. Practicals are used to investigate the validity of Hooke's law and to validate predictions about the natural frequencies and time periods of the various systems.

Chapter 5

This chapter extends the study of oscillations to include the pendulum and damped and forced oscillations in various contexts. Second order differential equations are derived and solved analytically, using complementary functions and particular integrals. The full pendulum equation is solved numerically and its solution compared to that for the simple pendulum.

Tasksheets

1 Modelling resisted motion

1.2   Resistance to motion

> (a)   Try dropping a ball of paper and a sheet of paper at the same time. Which falls faster? Account for your observations.
>
> (b)   Imagine holding your hand out of the window of a moving car. Describe what it would feel like.
>
> (c)   What is the force resisting motion in each of (a) and (b)?
>
> (d)   What two factors seem to affect the magnitude of this force?

(a)   The ball of paper falls to the ground more quickly. The sheet of paper tends to glide from side to side.

Assuming the ball of paper is made from a single sheet, then both objects have the same weight acting vertically downwards. However, the greater cross-sectional area of the sheet of paper means that there is a greater resistance to its motion through the air.

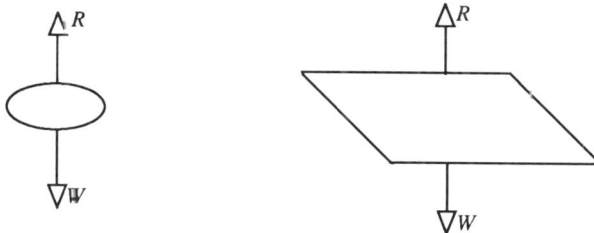

(b)   At high speed you would feel a substantial force pushing back on your hand. The force increases as the car's speed increases.

(c)   The force is called air resistance or drag and arises from movement through the air.

(d)   The drag force increases as either the area of the object or its speed increases.

Sky-diving

1.

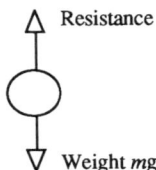

Resistance

Weight mg

2.   The sky-diver jumps and accelerates due to her weight. As her speed increases the resistance force increases and so the resultant force acting on her decreases. At some point, the upward resistance force will balance her weight and though she continues to fall, her acceleration is zero. At this speed the forces are balanced; this is her maximum speed, called the 'terminal speed'.

3.   Her maximum speed occurs when the resistance force equals her weight. To increase her maximum speed she must reduce the resistance force. She can do this by changing her posture, so that a smaller area is facing the oncoming air. To reduce her speed she must increase her area, for example by opening the parachute.

4.   As speed increases, the air resistance force increases.

5.   Two possibilities are:

(a) $R = Kv$, for some constant K

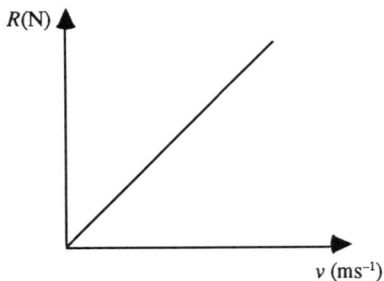

In this case the relationship is linear.

(b) $R = kv^2$, for some constant k

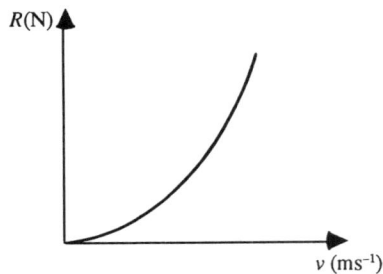

In this case the relationship is quadratic.

Free-fall

1.   Assumptions are:
  - the sky-diver is a particle of mass 60 kg;
  - her terminal speed is 50 ms^{-1};
  - gravity is constant and g = 10 ms^{-2};
  - let v ms^{-1} be her speed and h metres the height at time t.

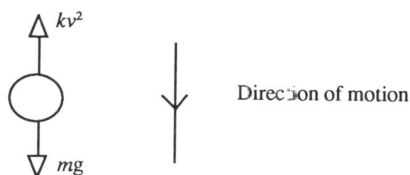

Direction of motion

Using Newton's second law downwards:

$$mg - kv^2 = m\, \frac{dv}{dt}$$

At terminal speed, $mg = kv^2 \Rightarrow k = \frac{mg}{v^2}$.

Using $m = 60$, g $= 10$ and $v = 50$ gives $k = 0.24$.

Substituting values into the equation of motion:

$$60\, \frac{dv}{dt} = 600 - 0.24v^2$$

$$\Rightarrow \frac{dv}{dt} = 10 - 0.004v^2$$

2.

Time t_n	Velocity v_n	Acceleration a_n	Height h_n
0	0	10	3700
0.1	1.00	10.000	3700.0
0.2	2.00	9.996	3699.9
0.3	3.00	9.984	3699.7
0.4	4.00	9.964	3699.4
0.5	5.00	9.936	3699.0
0.6	5.99	9.900	3698.5
0.7	6.98	9.857	3697.9
0.8	7.96	9.805	3697.2
0.9	8.94	9.746	3696.4
1.0	9.92	9.680	3695.5
...	....	...	...

(continued)

3.

Time t_n	Velocity v_n	Acceleration a_n	Height h_n
2.0	19.19	8.66	3681.3
3.0	27.21	7.20	3658.4
4.0	33.69	5.61	3628.2
5.0	38.64	4.16	3592.1
...	...	...	...
10.0	48.48	0.62	3367.5
15.0	49.82	0.08	3120.8
20.0	49.98	0.01	2871.2
25.0	50.00	0	2621.2
30.0	50.00	0	2371.2
35.0	50.00	0	2121.2
40.0	50.00	0	1871.2
45.0	50.00	0	1621.2
50.0	50.00	0	1371.2
55.0	50	0	1121.2
60.0	50	0	871.2
65.0	50	0	621.2
65.2	50	0	611.2

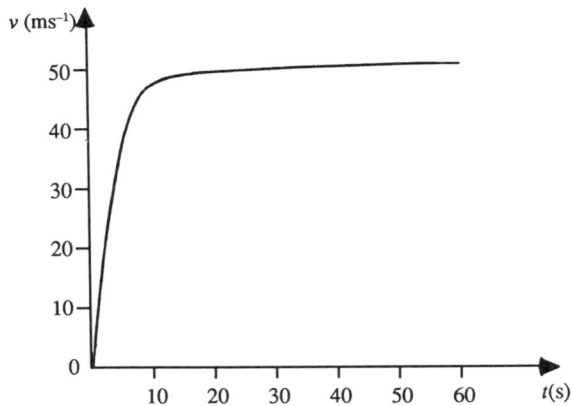

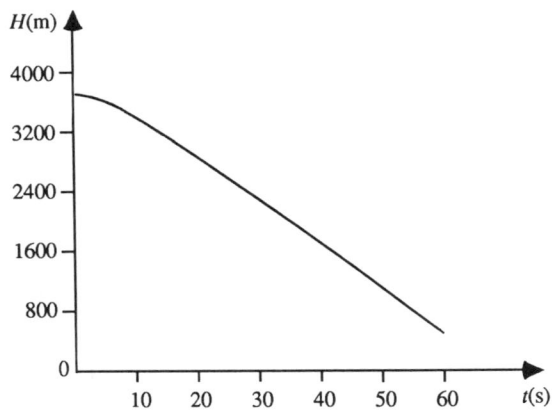

(continued)

4.   The sky-diver is almost at terminal speed from about 15 seconds onwards.

5.   The parachute is opened at 610 metres so the sky-diver has had about 65 seconds of free-fall.

6.   In the article, the free-fall time is at most one minute long, which is just under the value calculated. In the model, a terminal speed of 50 ms^{-1} was used, which is slightly slower than the speed in the text. Using this higher speed in the calculation would give a shorter free-fall time.

       The time to reach terminal speed is given as approximately 8 seconds. In the model, terminal speed was not reached until after approximately 15 seconds. However, after 8 seconds the sky-diver is falling at 46.5 ms^{-1} which is approaching her terminal speed.

7E.   A smaller step size will increase the accuracy. As the step size tends to zero, the solution should tend to a limit.

Resistance experiments

1 & 2   –

3.   You should find that the object reaches its terminal speed very quickly. The graph of distance against time is therefore linear for most of the distance fallen.

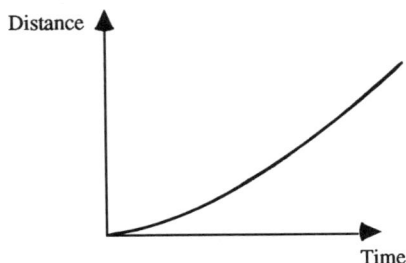

The terminal speed is the gradient of the straight-line section (where the object's speed is constant).

4.   As the mass increases, other things being equal, the terminal speed will increase. It may be necessary to drop the object from a considerable height for it to reach terminal speed.

5.   A linear graph for w against m suggests the $R = Kv$ model for resistance.

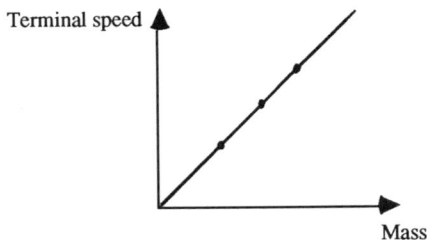

A graph such as the one below suggests that $w \propto \sqrt{m}$ and therefore indicates that the $R = kv^2$ model for resistance is appropriate.

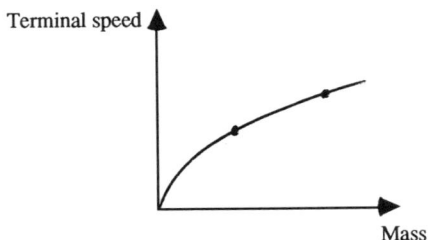

6.   You can use your own data to find an appropriate value for K or k.

1. (a) and (b)

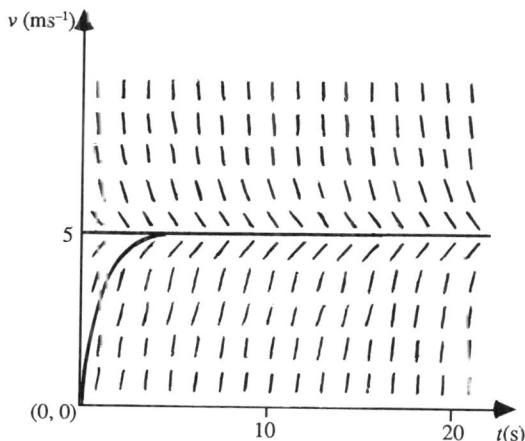

(c) The time taken to reach 4.5 ms^{-1} is 2.3 seconds. This can be found approximately from the graph or can be obtained analytically as follows.

The equation of motion is $\dfrac{dv}{dt} = 5 - v$

Separating variables gives $\displaystyle\int_{0}^{4.5} \frac{1}{5-v}\, dv = \int_{0}^{t} dt$

$$\Rightarrow \left[-\ln |5 - v| \right]_{0}^{4.5} = t$$

$$\Rightarrow \quad -\ln 0.5 + \ln 5 = t$$

$$\Rightarrow \quad\quad\quad t \approx 2.3$$

2. (a)

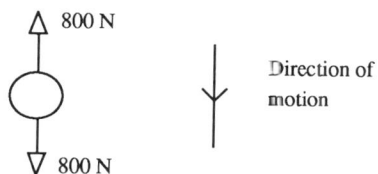

800 N

Direction of motion

800 N

(b) The area is halved so the resistance is halved. The new resistance force is 400 N.

(c) Newton's second law downwards:

$$800 - 400 = 80a \Rightarrow a = 5$$

The initial acceleration is 5 ms^{-2} downwards.

(continued)

13

3.　(a)　The speed of the car relative to the headwind is $(v + 10)$.

So the equation of motion is now $T - K(v + 10) = m \frac{dv}{dt}$.

(b)　The car's maximum speed will be 10 ms^{-1} less than it would be in still air.

(c)　The car in the slipstream experiences reduced air resistance. Consequently, it is possible to accelerate in the slipstream to a greater speed than the car in front. The rear car then can pull out and overtake before the air resistance slows it down.

4.　(a)

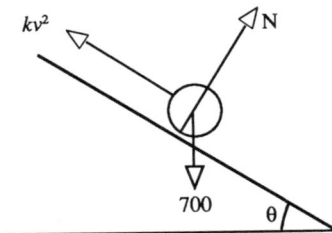

Using Newton's second law down the slope:

$$700 \sin \theta - kv^2 = 70 \frac{dv}{dt}$$

(b)　At maximum speed,　　$700 \sin \theta = kV^2$

$$\Rightarrow V = \sqrt{\left(\frac{700 \sin \theta}{k}\right)}$$

(c)　As θ increases, $\sin \theta$ increases. So as the slope becomes steeper, the maximum speed of the skier increases.

2 Analytical methods

2.1 Motion at 'low speeds'

(a) Interpret these graphs, referring to the acceleration and terminal speed of the bob.

(b) By integrating the expression for v, show that the distance fallen, x, is given by:

$$x = \frac{mg}{K}\left(t + \frac{m}{K}\left(e^{\frac{-Kt}{m}} - 1\right)\right)$$

(c) Draw graphs of distance against time for identically-shaped bobs of masses 1, 2, 5 and 10 kg, where $K = 5$.

(a) All the graphs show that the speed of the bob increases as the bob falls. The acceleration decreases as the bob approaches a terminal speed. The heavier the bob the greater its terminal speed; the longer it takes for this to be reached. (For example, the 10 kg bob has not yet reached terminal speed after 2.0 seconds.) Initially the bobs have the same acceleration.

(b) $\dfrac{dx}{dt} = \dfrac{mg}{K}\left(1 - e^{-\frac{Kt}{m}}\right)$

$x = \dfrac{mg}{K}\int\left(1 - e^{-\frac{Kt}{m}}\right)dt \Rightarrow x = \dfrac{mg}{K}\left(t + \dfrac{m}{K}e^{-\frac{Kt}{m}} + C\right)$

The initial conditions were $x = 0$, $t = 0$, so $C = -\dfrac{m}{K}$.

Then, $x = \dfrac{mg}{K}\left(t + \dfrac{m}{K}e^{-\frac{Kt}{m}} - \dfrac{m}{K}\right) \Rightarrow x = \dfrac{mg}{K}\left(t + \dfrac{m}{K}\left(e^{-\frac{Kt}{m}} - 1\right)\right)$

(c)

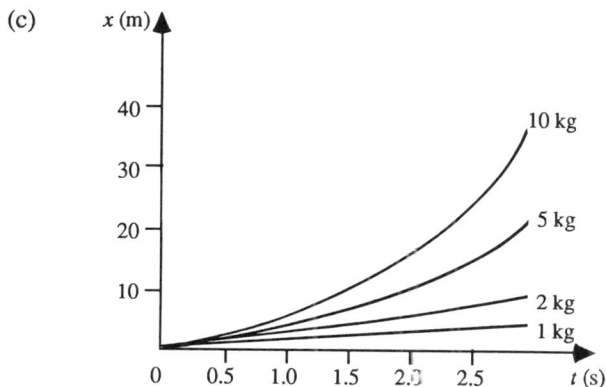

A falling feather

1.

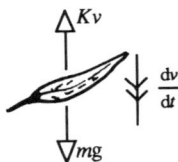

By Newton's second law:

$$m\frac{dv}{dt} = mg - Kv$$

At terminal speed, $Kw = mg$

$$\Rightarrow K = \frac{mg}{w}$$

2. $$m\frac{dv}{dt} = mg - Kv \Rightarrow m\frac{dv}{dt} = mg - \frac{mg}{w}v$$

$$\Rightarrow \frac{dv}{dt} = g\left(1 - \frac{v}{w}\right)$$

3. (a) $v < w$ The feather is falling more slowly than terminal speed. However, $\frac{dv}{dt} > 0$ so the feather is accelerating downwards. As its speed increases the acceleration decreases to zero.

    (b) $v = w$ The feather is falling at terminal speed and so there is no acceleration.

    (c) $v > w$ The feather is moving faster than its terminal speed. Since $\frac{dv}{dt} < 0$ the feather is decelerating. It will slow down until it is moving at its terminal speed.

4. $$\frac{dv}{dt} = \frac{g}{w}(w - v)$$

    Separating variables, $$\int_0^v \frac{1}{w-v}\,dv = \int_0^t \frac{g}{w}\,dt$$

    $$\Rightarrow -\ln|w - v| + \ln|w| = \frac{gt}{w}$$

    $$\Rightarrow \ln\left|\frac{w-v}{w}\right| = -\frac{gt}{w}$$

    $$\Rightarrow \frac{w-v}{w} = e^{-\frac{gt}{w}}$$

    $$\Rightarrow v = w\left(1 - e^{-\frac{gt}{w}}\right)$$

(continued)

16

5. As $t \to +\infty$, then the limit of $e^{-\frac{gt}{w}}$ is 0. Thus the limit of the feather's speed, v, as $t \to \infty$ is the terminal speed w. However, this is never achieved exactly in finite time.

6.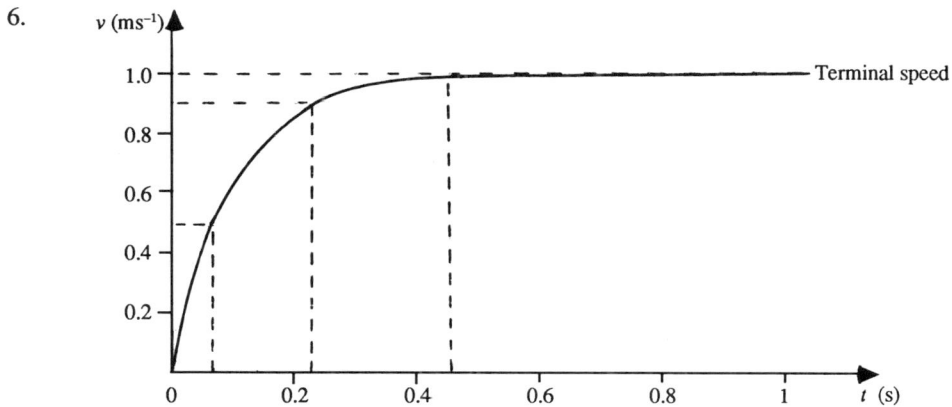

    The time taken to reach:

    (a) 50% is 0.07 second;
    (b) 90% is 0.23 second;
    (c) 99% is 0.46 second.

7. The results show how quickly the feather approaches terminal speed. In less than 0.5 second it has reached 99% of its terminal speed. However, this is twice the time taken to reach 90%.

    In practice, it is not reasonable to discuss the speed of a feather to very great accuracy since it will vary considerably according to air disturbances not accounted for in this simple model.

8. (a) $\dfrac{dv}{dt} = \dfrac{dv}{dx} \times \dfrac{dx}{dt}$, by the chain rule

    $\Rightarrow \dfrac{dv}{dt} = \dfrac{dv}{dx} \times v = v\dfrac{dv}{dx}$

    (b) $v\dfrac{dv}{dx} = g\left(1 - \dfrac{v}{w}\right)$

(continued)

9. $$\frac{v}{v-w} = \frac{v-w}{v-w} + \frac{w}{v-w} = 1 + \frac{w}{v-w}$$

$$v\frac{dv}{dx} = \frac{-g}{w}(v-w)$$

$$\Rightarrow \int \frac{v}{v-w}\ dv = \int \frac{-g}{w}\ dx$$

$$\Rightarrow \int \left(1 + \frac{w}{v-w}\right) dv = -\frac{gx}{w} + C$$

$$\Rightarrow v + w\ \ln|v-w| = -\frac{gx}{w} + C$$

Initial conditions are $v = 0,\ x = 0 \Rightarrow C = w\ \ln|w|$

Then $v + w\ \ln|v-w| = -\frac{gx}{w} + w\ \ln|w|$

$$\Rightarrow v + w\ \ln\left|\frac{v-w}{w}\right| = -\frac{gx}{w}$$

$$\Rightarrow x = \frac{-w}{g}\left(v + w\ \ln\left|\frac{v-w}{w}\right|\right)$$

10.

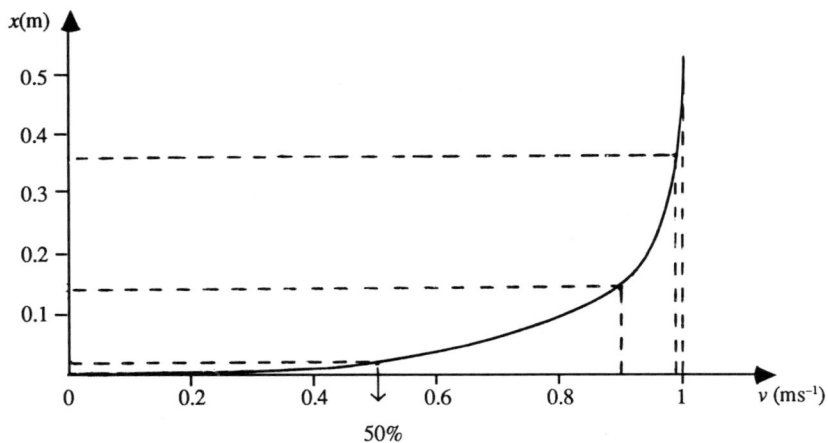

50%

11. From the graph, the distance taken to reach:

   (a) 50% of terminal speed is 0.02 m;
   (b) 90% of terminal speed is 0.14 m;
   (c) 99% of terminal speed is 0.36 m.

Opening the parachute

1.   Possible assumptions are:

- the sky-diver is a particle of mass 60 kilograms;
- there is no wind;
- the sky-diver initially is travelling with speed 50 ms^{-1};
- the parachute opens instantaneously;
- the terminal speed with the parachute is 5 ms^{-1};
- a safe landing speed is 5.5 ms^{-1} (equivalent to jumping from a height of 1.5 metres).

2.

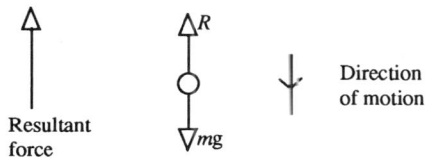

Resultant force R Direction of motion mg

3.   At terminal speed, $mg = kW^2 \Rightarrow k = \dfrac{mg}{W^2}$

Then
$$mg - mg\,\frac{v^2}{W^2} = mv\,\frac{dv}{dx}$$

$$\Rightarrow \int_U^v \frac{v}{g\left(1 - \frac{v^2}{W^2}\right)}\,dv = \int_0^x dx$$

$$\Rightarrow \frac{W^2}{g}\int_U^v \frac{v}{W^2 - v^2}\,dv = \frac{-W^2}{2g}\left[\ln\left|W^2 - v^2\right|\right]_U^v = x$$

$$\Rightarrow x = \frac{-W^2}{2g}\ln\left|\frac{W^2 - v^2}{W^2 - U^2}\right|$$

4.   For a safe landing speed of V, the distance fallen is

$$H = \frac{-W^2}{2g}\ln\left|\frac{W^2 - V^2}{W^2 - U^2}\right|$$

Using the values $W = 5$, $V = 5.5$, $g = 10$ and $U = 50$, the least height at which the parachute can be opened is found to be 7.7 metres.

5.   The value obtained is considerably lower than that suggested as a safe height. In practice, time must be allowed for the parachute to open. Furthermore, there must be a safety margin to allow time for the first parachute to fail to open and for a reserve parachute to be used. At a height of 610 m the sky-diver has only 12 seconds before impact when travelling at 50 ms^{-1}.

1. (a)

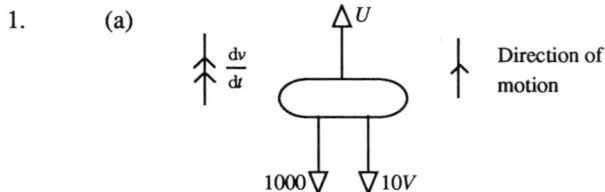

By Newton's second law:

$$U - 1000 - 10v = 100 \frac{dv}{dt}$$

At terminal speed w, $10w = U - 1000$.

$$\Rightarrow w = \frac{U - 1000}{10}$$

The terminal speed increases as U increases. However, if $U = 1000$ then the terminal speed is zero and the diver will float without rising or sinking. (The diver is said to be **neutrally buoyant**). If $U < 1000$ the diver will sink to the bottom.

(b) Replacing $U - 1000$ by $10w$ in the equation of motion,

$$10w - 10v = 100 \frac{dv}{dt}$$

$$\Rightarrow 10 \frac{dv}{dt} = w - v$$

(c) Separating variables gives:

$$\int_0^{0.9w} \frac{10}{w - v} \, dv = \int_0^t dt$$

$$\Rightarrow \left[-10 \ln (w - v) \right]_0^{0.9w} = t$$

$$\Rightarrow t = -10 \ln (0.1w) + 10 \ln (w) = 10 \ln 10$$

$$\Rightarrow t \approx 23$$

The diver reaches 90% of terminal speed after 23 seconds.

(continued)

2. (a) The main assumptions are:

- the upthrust force U newtons is constant;
- the resistance force is modelled as v^2 newtons;
- the motion is in a vertical line (or alternatively v is the vertical component of the velocity);
- there is no wind, or airstream;
- the weight is constant.

(b) Using $v\dfrac{dv}{dx}$ for acceleration, the equation of motion becomes

$$mv\,\frac{dv}{dx} = (10m - U) - v^2$$

(c) Substituting for $U = 12000$ and $m = 1500$

$$1500\,v\,\frac{dv}{dx} = 3000 - v^2$$

$$\Rightarrow\ 1500 \int_0^v \frac{v}{3000 - v^2}\,dv = \int_0^{1000} dx$$

$$\Rightarrow\ -750 \left[\ \ln \mid 3000\ v^2\mid\ \right]_0^v = 1000$$

$$\Rightarrow\ \ln \left|\frac{3000 - v^2}{3000}\right| = -\frac{4}{3}$$

$$\Rightarrow\ 3000 - v^2 = 3000e^{-\frac{4}{3}}$$

$$v \approx 47,\ \text{so the speed of impact is 47 ms}^{-1}.$$

3. (a) $\dfrac{dx}{dt} = 12\left(1 - e^{-\frac{5t}{6}}\right)$

The terminal speed of the ball (as $t \to +\infty$) is $12\ \text{ms}^{-1}$.

(b) The distance fallen by the second ball is:

$$Y = 12\left[T + 1.2\left(e^{-\frac{5T}{6}} - 1\right)\right]\ \text{where } T = t - 1$$

The distance between the two balls is:

$$12\left[t - (t-1) + 1.2\left(e^{-\frac{5t}{5}} - e^{-\frac{5(t-1)}{6}}\right)\right] = 12\left[1 + 1.2\,e^{-\frac{5t}{6}}\left(1 - e^{\frac{5}{6}}\right)\right]$$

When $t = 1$, $12\left[1 + 1.2\,e^{-\frac{5}{6}}\left(1 - e^{\frac{5}{6}}\right)\right] = 3.85$

Initially, the balls are 3.85 metres apart. As they fall this distance increases to 12 metres apart when the balls are travelling at their terminal speeds. (This will depend on how deep the well is!)

(continued)

21

4.   (a)

900 △ △ 225v

↓ Direction of motion

▽
750

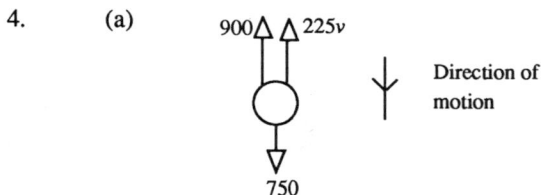

By Newton's second law:

$$750 - 900 - 225v = 75v \frac{dv}{dx}$$

$$\Rightarrow -150 - 225v = 75v \frac{dv}{dx}$$

$$\Rightarrow \qquad v \frac{dv}{dx} = -3v - 2$$

(b)   Then $\displaystyle\int_{15}^{0} \frac{v}{3v+2}\, dv = \int_{0}^{d} -1 \, dx$, where d metres is the greatest depth reached.

$$\frac{1}{3}\int_{15}^{0} \left(1 - \frac{2}{3v+2}\right) dv = \frac{1}{3}\left[v - \frac{2}{3}\ln(3v+2)\right]_{15}^{0} = -d$$

$$\Rightarrow -15 - \frac{2}{3}\ln\left(\frac{2}{47}\right) = -3d$$

$$\Rightarrow d = 5 + \frac{2}{9}\ln\left(\frac{2}{47}\right) = 4.3$$

The greatest depth reached is 4.3 metres.

- The assumptions made to obtain the equation of motion ignored the height of the diver. Since the depth reached is only 4.3 metres, the height is obviously significant.

- The resistance force has been modelled as 225v, whereas $R = kv^2$ may be more appropriate because of the speed at which the diver is travelling.

- Additionally, some speed would be lost as the diver enters the water.

3 *Variable mass and weight*

3.2 Escaping from the Earth

> **Discuss the factors involved in modelling the launch of a rocket into space.**

There are a number of factors that could be introduced to a model. The more obvious ones are listed here.

Air resistance   As the rocket moves higher and higher in the Earth's atmosphere and beyond, the resistance force will decrease.

Direction   The first stage of the launch is usually to put the rocket into orbit about the Earth. From here, a fairly small amount of energy is needed to escape into space. It may be easier to consider a rocket as travelling directly away from the Earth, thus ignoring the trajectory required for orbit.

Gravity   At the launch itself the distance travelled is relatively small so that gravity could be modelled as a constant. However, for motion into outer space the distances are so great that the variation in gravity would be significant.

Other bodies   Once the rocket is far away from the Earth the gravitational attraction of other planets will become more significant than the attraction of the Earth. At some point, the rocket may be subject to significant attraction of two or even three bodies, such as the Moon, the Earth and the Sun.

Earth's rotation   Since the rocket is travelling great distances over a rotating Earth it is subject to a force arising from the actual rotation. This force is known as the Coriolis force and will alter the path of the rocket slightly. To simplify matters this force will be ignored.

Rocket   A Saturn V rocket was used in the Apollo space missions. This huge rocket consisted of three stages with a lunar module and a command module at the top of the stack. In total, the Saturn V rocket was 363 feet tall with a mass of about 3000 tonnes.

Fuel   The first stage of the Saturn V rocket burnt about 2100 tonnes of fuel in the first 160 seconds. The second stage burnt 450 tonnes in 6.5 minutes. In the final stage 120 tonnes of fuel were burnt in two separate phases, the first (lasting 150 seconds) to steer the rocket into Earth orbit and the second (lasting 345 seconds) altered the course to head towards the Moon.

Thrust

The initial thrust produced by the engines must be greater than the weight of the rocket or it would not take off.

Estimates of the thrust produced by each stage of the Saturn V rocket are:

- for the first stage, five F-1 engines producing a total of 33 million newtons of thrust;

- for the second stage, five J-2 engines producing a total of about 4.4 million newtons of thrust;

- for the third stage, a single J-2 engine producing about 0.9 million newtons of thrust.

In the following sections some of these factors will be considered to provide suitable models.

The Moon-lander

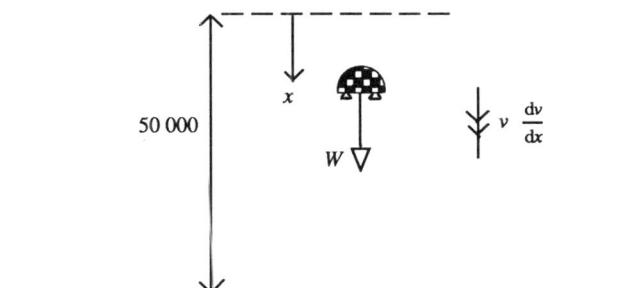

1. Using Newton's law of gravitation and second law towards the surface:

    $$mv\frac{dv}{dx} = \frac{GMm}{r^2}$$

    where r is the distance of the lander from the centre of mass of the Moon.

    So $r = R + 50\,000 - x$ where R is the radius of the Moon.

    Substituting for r :

    $$mv\frac{dv}{dx} = \frac{GMm}{(R + 50\,000 - x)^2}$$

2. Separating variables and integrating:

    $$\int_0^v v\,dv = \int_0^x \frac{GM}{(R + 50\,000 - x)^2}\,dx$$

    $$\Rightarrow \frac{1}{2}v^2 = \left[\frac{GM}{(R + 50\,000 - x)}\right]_0^x = GM\left(\frac{1}{(R + 50\,000 - x)} - \frac{1}{(R + 50\,000)}\right)$$

    $$\Rightarrow v^2 = \frac{2GM\,x}{(R + 50\,000 - x)(R + 50\,000)}$$

    $$\Rightarrow v = \sqrt{\left(\frac{2GMx}{(R + 50\,000 - x)(R + 50\,000)}\right)}$$

3. When $x = 50\,000$, $v = \sqrt{\left(\frac{2GM \times 50\,000}{R(R + 50\,000)}\right)} \approx 397$

    The impact speed is 397 ms^{-1}.

(continued)

25

4. If gravity is modelled using Newton's law of gravitation, the initial acceleration of the lander is smaller, but this increases as the lander moves closer to the surface.

    However, the impact speed found using the constant gravity model is 402.5 ms^{-1}, which is only slightly faster. For motion from a height of 50 km the two models give solutions that are similar.

5. If the initial height is h, then the speed after falling x metres becomes:

    $$v = \sqrt{\left(\frac{2GMx}{(R+h-x)(R+h)}\right)}$$

    On impact, the distance fallen $x = h$, so the impact speed is:

    $$V_h = \sqrt{\left(\frac{2GMh}{R(R+h)}\right)}$$

6. (a)

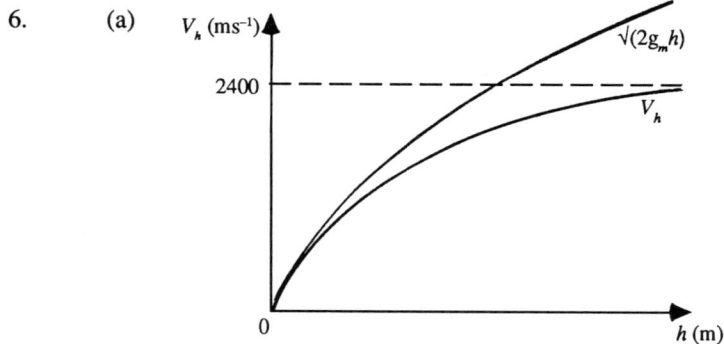

    When the lander is at a very great distance from the Moon, its weight is very small, almost negligible. As the lander falls toward the Moon, it initially accelerates very slowly.

    As the lander gets nearer the Moon, its weight increases and it begins to accelerate more rapidly.

    From the graph, it is clear there is a limit to the impact speed, so that falling from greater heights will not increase the impact speed.

    (b) The superimposed graph shows that $v = \sqrt{(2g_m h)}$ (for $g_m = 1.62$) is a good approximation for small h.

7E. $g_m = \dfrac{GM}{R^2} \Rightarrow V_h = \sqrt{\left(\dfrac{2g_m hR}{R+h}\right)} = \left(1 + \dfrac{h}{R}\right)^{-\frac{1}{2}} \sqrt{(2g_m h)}$

    $\Rightarrow V_h \approx \left(1 - \dfrac{h}{2R}\right) \sqrt{(2g_m h)}$ if $\dfrac{h}{R}$ is small.

Rocket propulsion

1. In theory you and the trolley should roll the other way, though this will depend on the friction in the wheel bearings.

2. (a) The greater the speed of the balls the more momentum they have, and since momentum is conserved the trolley will move the opposite way with greater speed.

   (b) More massive balls have greater momentum and the effect on the trolley will be the same as in (a) above.

   (c) Each time a ball is thrown there is an exchange of momentum and the mass of the trolley with the remaining balls decreases. The rate at which the balls are thrown affects the acceleration of the trolley; the quicker the rate, the greater the acceleration.

3.
Before   After

$0 \leftarrow \boxed{60 + 198 + 2}$ $v_1 \text{ ms}^{-1} \leftarrow \boxed{60 + 198}$ $\boxed{2}\rightarrow 10 \text{ ms}^{-1}$

4. Applying conservation of momentum:

$$0 = (60 + 198)v_1 - 20$$
$$\Rightarrow v_1 = \frac{10}{129}$$

The speed of the trolley is $\frac{10}{129}$ ms^{-1}.

5.
Before   After

$v_1 \text{ ms}^{-1} \leftarrow \boxed{60 + 198}$ $v_2 \text{ ms}^{-1} \leftarrow \boxed{60 + 196}$ $\boxed{2}\rightarrow 10$

$$258 \times \frac{10}{129} = 256v_2 + 2\left(\frac{10}{129} - 10\right)$$
$$\Rightarrow v_2 = 0.156 \text{ ms}^{-1}$$

$$256 \times 0.156 = 254 v_3 + 2(0.156 - 10)$$
$$\Rightarrow v_3 = 0.234 \text{ ms}^{-1}$$

$$254 \times 0.234 = 252 v_4 + 2(0.234 - 10)$$
$$\Rightarrow v_4 = 0.314 \text{ ms}^{-1}$$

(continued)

6.   (a)   Applying the principle of conservation of momentum:

$$(M + 100m)\,v_0 = (M + 99m)\,v_1 + m\,(v_0 - C)$$

$$\Rightarrow \quad (M + 99m)\,v_0 = (M + 99m)\,v_1 - mC$$

$$\Rightarrow \quad v_0 = v_1 - \frac{mC}{M + 99m}$$

$$\Rightarrow \quad v_1 = v_0 + \frac{mC}{M + 99m}$$

(b) $v_2 = v_1 + \dfrac{mC}{M + 98m}$

(c) $v_n = v_{n-1} + \dfrac{mC}{M + (100 - n)m}$

7.   Substituting $m = 2$, $M = 60$ and $C = 10$ into the relations given in question 6, gives solutions which are the same as those obtained in question 5.

$$v_0 = 0 \text{ ms}^{-1}, \quad v_1 = 0.078 \text{ ms}^{-1}, \quad v_2 = 0.156 \text{ ms}^{-1}, \quad v_3 = 0.234 \text{ ms}^{-1}, \quad v_4 = 0.314 \text{ ms}^{-1}$$

8.   Using a program, the following speeds after each set of ten throws are obtained by substituting $m = 2$, $M = 60$ and $C = 10$ into the sum:

$$v_n = \frac{20}{60 + 2\,(100 - n)} + \ldots + \frac{20}{60 + 2 \times 97} + \frac{20}{60 + 2 \times 98} + \frac{20}{60 + 2 \times 99}$$

Balls thrown	0	10	20	30	40	50	60	70	80	90	100
Speed (ms⁻¹)	0	0.80	1.68	2.64	3.70	4.88	6.22	7.78	9.62	11.88	14.80

The final speed of the trolley is 14.80 ms⁻¹.

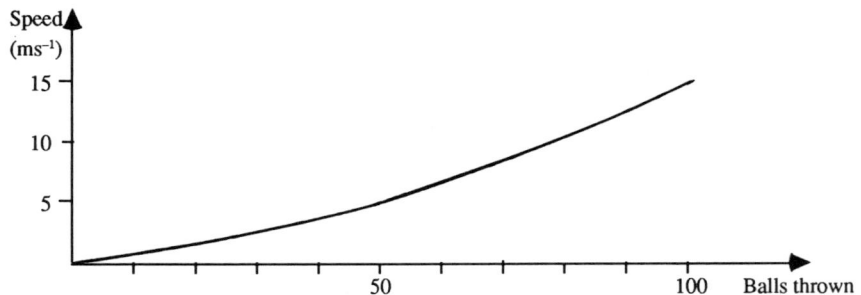

From the graph, you can see that the acceleration increases with time. As more balls are thrown from the trolley, the actual mass of you, the trolley and the remaining balls is decreasing. The increase in momentum due to an ejected ball results in a greater increase in the trolley's speed as it is becoming lighter.

9E.   Increasing any of these three variables increases the rate of change of the momentum of the ejected mass and hence the rate of change of the forward momentum of the trolley.

28

Modelling rocket motion

1. (a) $(M - \mu dt)(v + dv) + \mu dt\,(v - C) = Mv + M\,dv - \mu C\,dt - \mu dv\,dt$

    (b) Change in momentum $= M\,dv - \mu C\,dt - \mu dv\,dt$

    $$\frac{\text{Change of momentum}}{dt} = M\frac{dv}{dt} - \mu C - \mu dv$$

    As $dt \to 0$, $dv \to 0$ and so

    Rate of change of momentum $= M\frac{dv}{dt} - \mu C$

2. The only external force is the total weight of the rocket and fuel. This is Mg downwards throughout the time interval.

    Applying Newton's second law:

    $$-Mg = M\frac{dv}{dt} - \mu C$$

    $$\Rightarrow M\frac{dv}{dt} = \mu C - Mg$$

3. Applying Newton's second law.

    $$-T - \mu\,dt\,g \approx \frac{\mu\,dt\,(v-C) - \mu\,dt\,v}{dt} = -\mu C$$

    Let $dt \to 0$, then $T = \mu C$.

4. The thrust must be greater than Mg, i.e. $\mu C > Mg$.

5E. Lift-off can only occur when $M = \dfrac{\mu C}{g}$

    $$\Rightarrow M_0 - \mu\tau = \frac{\mu C}{g}$$

    $$\Rightarrow \qquad \tau = \frac{M_0}{\mu} - \frac{C}{g}$$

    (a) When $\tau = 0$, $\mu C = gM_0$.

    The rocket lifts off immediately. The rate at which fuel is burnt and the exhaust speed just provide enough thrust to balance the initial weight of the rocket.

    (b) When $\tau > 0$, $\mu C < gM_0$.

    The rocket's weight is greater than the thrust. If the rocket is on the launch pad then there must be an additional reaction force R on the rocket such that $R + T - Mg = 0$. Time is taken while fuel is burnt and ejected, decreasing the weight of the rocket until R is zero and the thrust is sufficient for lift-off.

    In practice, the exhaust speed and the rate of burn take time to reach their maximum, i.e. μ and C are not constant initially.

1.	(a)	At $t = 0$, $\frac{dv}{dt} = -10 + \frac{1250}{100} = 2.5$

Initial acceleration is 2.5 ms^{-2} upwards.

(b)

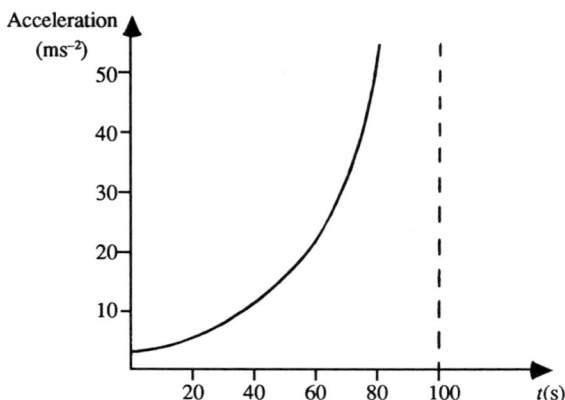

(c)	$\frac{dv}{dt} = -10 + \frac{1250}{100-t}$

Integrating:	$v = \int_{0}^{80} \left(-10 + \frac{1250}{100-t}\right) dt$

$\Rightarrow\ v = \left[-10t - 1250 \ln \left| 100 - t \right|\right]_{0}^{80}$

$\Rightarrow\ v = -800 - 1250 \ln \left| \frac{20}{100} \right| = 1211.8$

The speed of the rocket when the fuel runs out is approximately 1210 ms^{-1}.

(d)	This model would give an over-estimate because:

· air resistance has been ignored;
· the rocket might not travel vertically for the whole motion;
· the rate at which fuel is burnt may decrease as the fuel runs out.

(continued)

2. (a) (i) At $t = 30$ the first stage of the rocket has run out of fuel and is detached. At this point there is no thrust acting on the rocket so it slows down due to the action of its weight.

   (ii) At $t = 35$ the second stage begins its burn and the rocket accelerates again. The rate of acceleration is greater now, since the rocket has lost the mass of the first stage.

   (b) For the second stage, $10\ 000 \dfrac{dv}{dt} = -100\ 000 + 100C$

   For the third stage, $5800 \dfrac{dv}{dt} = -58\ 000 + 100C$

3. (a) Inital mass of module is 1800 kg.

   Mass of module at time t, $M = 1800 - 50t$

   (b) Substituting into the equation of motion:

   $$(1800 - 50t)\frac{dv}{dt} = (1800 - 50t)g - 18\ 000$$

   $$\frac{dv}{dt} = 1.62 - \frac{1800}{180 - 5t}$$

   Integrating: $$\int_{100}^{v} dv = \int_{0}^{t} \left(1.62 - \frac{360}{36 - t}\right) dt$$

   $$\Rightarrow \quad v - 100 = 1.62t + 360 \ln\left|\frac{36 - t}{36}\right|$$

   $$\Rightarrow \quad v = 100 + 1.62t + 360 \ln\left|\frac{36 - t}{36}\right|$$

   (c) The graph of descent speed against time is as shown:

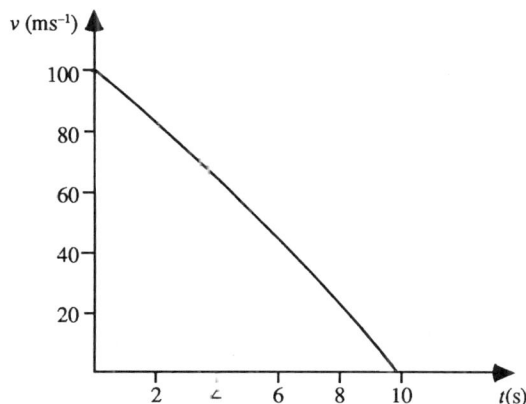

   When $v = 0$, from the graph $t = 9.92$ seconds.

   The lander takes 9.92 seconds to slow to zero descent.

   (continued)

4. (a)

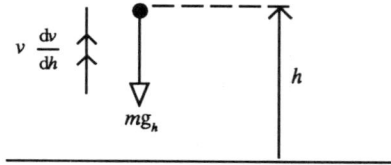

Let m kilograms be the mass of the projectile.

Using Newton's second law away from the Moon:

$$mv\,\frac{dv}{dh} = -m\,g_h = -m \times 1.62\left(1 - \frac{2h}{r}\right)$$

$$\Rightarrow mv\,\frac{dv}{dh} = -1.62\left(1 - \frac{2h}{R}\right)$$

(b) Separating variables and integrating:

$$\int_{U}^{0} v\,dv = -1.62\int_{0}^{H}\left(1 - \frac{2h}{R}\right)dh \text{ , where } H \text{ metres is the maximum height.}$$

$$\Rightarrow -\frac{1}{2}U^2 = -1.62\left(H - \frac{H^2}{R}\right)$$

$$\Rightarrow U = \sqrt{\left(3.24\left(H - \frac{H^2}{R}\right)\right)}$$

(c) The model for g_h is valid if the launch speed is not too great and the rocket's maximum height is much less than R.

i.e. for $\frac{H}{R}$ very small, $U \approx \sqrt{(3.24H)}$

However, as the launch speed increases and the maximum height approaches R the model is no longer valid. For such distances, Newton's law of gravitation is the most appropriate model.

When $H = R$ then $U = 0$ which is clearly incorrect. Hence the model is not valid.

This model would not be appropriate to calculate the escape speed as, in this case, $H \rightarrow +\infty$.

4 *Simple harmonic motion*

4.3   Modelling an oscillating body

> Using a spring or string with different loads, investigate the following questions.
>
> (a) What effect does the magnitude of the initial displacement have on the time period of oscillation?
>
> (b) For how many oscillations can the amplitude be assumed to be constant?
>
> (c) Comment on the validity of Hooke's law as a model for the tension in the spring or string throughout the motion. (Consider the case where the spring is in compression or where the string becomes slack.)
>
> (d) What other assumptions need to be made to set up a model?

(a)   Up to a point, the period appears to be independent of the initial displacement. The greater extension results in a larger restoring force, so the mass moves faster. This compensates for the greater distance to be travelled. However, if the displacement is too large the string may go slack.

(b)   If the initial amplitude is not too large then for some springs the amplitude remains reasonably constant for a large number of oscillations. Elastic strings are usually more erratic and the amplitude dies away in a much shorter time. However, the motion for strings and springs can become erratic if they are overloaded or the amplitude is too large.

(c)   For an elastic string, the tension, T, may be modelled as proportional to its extension, x,

$$T = kx$$

where k is the elastic constant of the string, which depends on the strength of the elastic, its cross-sectional area and its original length.

This relationship is known as Hooke's law. It is usually a valid assumption only for a limited range of extensions. The range depends on the string being used. It does **not** apply when $x < 0$, i.e. when the string becomes slack.

As with strings, the tension in a spring can also be modelled with Hooke's law. Again there is a limit to the amount a spring can be stretched without deforming the metal, beyond which the relationship no longer holds.

However, for many real springs a better model for the tension is $T = T_0 + kx$, where T_0 is the pre-tensioning force of the spring (i.e. the force that needs to be applied before the spring begins to stretch).

Some springs can be compressed. The thrust force in a compressed spring may also obey Hooke's law, though this is uncommon for real springs.

The graph shows possible relationships between the tension in a spring and the extension.

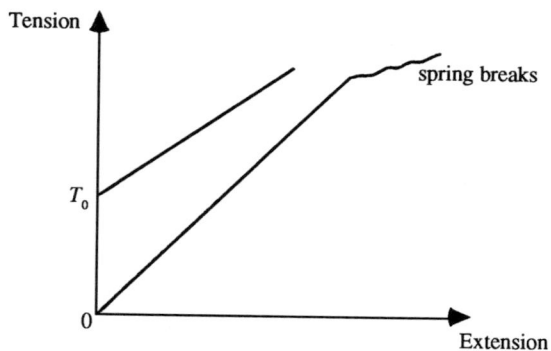

(d) Other assumptions may include:

- the baby is a particle of mass m kilograms;
- the baby is suspended from a spring so that it is never in contact with the floor;
- air resistance is negligible;
- the baby is given an initial displacement small enough for the spring never to become slack or compress.

Validating the formula $T \propto \sqrt{m}$

1.   Attach the spring to a support and measure the extension for different masses suspended in equilibrium. In equilibrium, the tension equals the weight suspended.

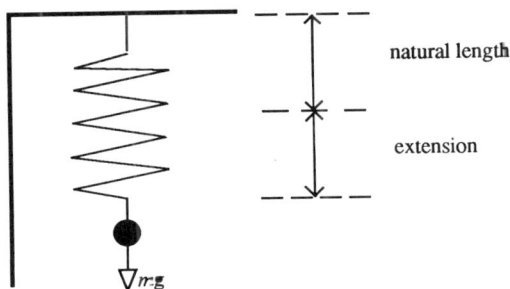

natural length

extension

Assuming Hooke's law is valid, then the spring constant is simply $k = \dfrac{\text{weight}}{\text{extension}}$.

Drawing a graph of weight against extension, an approximate value for k is the gradient of the line of best fit.

You will find that some springs are pre-tensioned and do not extend until a certain weight is applied. A better model for the tension in these springs would be $T = T_0 + kx$, where T_0 is the pre-tensioning force.

2.   Timing a number of oscillations, for example 5, 10 or 20, and calculating the average should give an accurate and reliable value for the time period of oscillation. Springs oscillate fairly consistently if the mass on the end is not too small or too great and you can assume that the oscillations are all equal. Decide on the initial amplitude of release to ensure consistency and repeatability.

3.   Collecting data for the time period for a range of masses will enable you to plot a graph of time period against mass. By fitting a curve to the data points you can compare the theoretical result that $T \propto \sqrt{m}$ (i.e $T = A\sqrt{m}$, where $A = \dfrac{2\pi}{\sqrt{k}}$) with your experimental result.

Any of the springs from the kit should give results that validate the theoretical result and it is possible to compare the value of k in the model for tension derived from the graph with the value obtained in the earlier experiment.

(An alternative approach is to time the period when the mass is, for example, 60 grams and then to check that the period is doubled when the mass is 240 grams.)

Solving the SHM equation

1. $v \dfrac{dv}{dx} + \omega^2 x = 0$

     Separating variables: $\int v \, dv = \int -\omega^2 x \, dx$

         $\Rightarrow \dfrac{1}{2} v^2 = -\omega^2 \dfrac{x^2}{2} + A$

         $\Rightarrow \quad v^2 = -\omega^2 x^2 + \text{constant}$

2.   With the initial condition $x = a$, $v = 0$:

         $0 = -\omega^2 a^2 + \text{constant}$

         $\Rightarrow \text{constant} = \omega^2 a^2$

     Substituting this in the solution: $v^2 = -\omega^2 x^2 + \omega^2 a^2 = \omega^2 (a^2 - x^2)$

         $\Rightarrow v = \pm \omega \sqrt{(a^2 - x^2)}$

3.   The graph of v against x from $x = -a$ to $x = a$ is as shown.

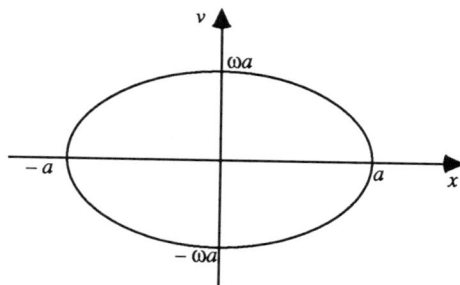

     The motion is such that when the displacement is at its maximum (or minimum) the velocity is zero, and when the displacement is zero the velocity is a maximum (or minimum).

4.   Writing $v = \dfrac{dx}{dt}$, $\dfrac{dx}{dt} = \pm \omega \sqrt{(a^2 - x^2)}$

     Separating variables: $\displaystyle\int_0^t dt = \int_a^x \dfrac{1}{\pm \omega \sqrt{(a^2 - x^2)}} \, dx$

     $\Rightarrow t = \left[\pm \dfrac{1}{\omega} \cos^{-1} \left(\dfrac{x}{a} \right) \right]_a^x = \pm \dfrac{1}{\omega} \left[\cos^{-1} \left(\dfrac{x}{a} \right) - \cos^{-1} 1 \right]$

     $\Rightarrow t = \pm \dfrac{1}{\omega} \cos^{-1} \left(\dfrac{x}{a} \right)$

     $\Rightarrow \dfrac{x}{a} = \cos (\pm \omega t) = \cos \omega t$

     $\Rightarrow x = a \cos \omega t$

                                                (continued)

5. For the mass on the spring, $\ddot{x} = -\frac{k}{m}x \Rightarrow v\frac{dv}{dx} = -\frac{k}{m}x$

Separating variables: $\int_{9}^{v} v\,dv = \int_{0}^{x} -\frac{k}{m}x\,dx$

$$\Rightarrow \frac{v^2}{2} - \frac{81}{2} = -\frac{k}{m}\frac{x^2}{2}$$

$$\Rightarrow v = \pm\sqrt{(81 - \frac{k}{m}x^2)}$$

Separating variables and integrating again: $\int_{0}^{t} dt = \int_{0}^{x} \frac{\pm 1}{\sqrt{(81 - \frac{kx^2}{m})}}\,dx$

$$\Rightarrow t = \pm\left[\sqrt{(\frac{m}{k})}\cos^{-1}(\sqrt{(\frac{k}{m})}\frac{x}{9})\right]_{0}^{x}$$

$$\Rightarrow t = \pm\sqrt{(\frac{m}{k})}\left(\cos^{-1}(\frac{x}{9}\sqrt{(\frac{k}{m})}) - \cos^{-1} 0\right)$$

$$\Rightarrow \cos^{-1}(\frac{x}{9}\sqrt{(\frac{k}{m})}) = \pm\sqrt{(\frac{k}{m})}t \pm \frac{\pi}{2}$$

$$\Rightarrow x = 9\sqrt{(\frac{m}{k})}\cos(\frac{\pi}{2} \pm \sqrt{(\frac{k}{m})}t)$$

Then $x = +9\sqrt{(\frac{m}{k})}\sin(\sqrt{(\frac{k}{m})}t)$ to satisfy the initial conditions.

These initial conditions may occur if the mass is struck from its equilibrium position with an initial speed of 9 ms^{-1}.

6. Let $x = Ae^{pt}$, then differentiating:

$\dot{x} = Ape^{pt}$ and $\ddot{x} = Ap^2e^{pt}$

Substituting into the equation of motion:

$Ap^2e^{pt} + \omega^2 Ae^{pt} = 0$

$\Rightarrow \quad p^2 + \omega^2 = 0$

7. Solving the auxillary equation, $p = \pm\sqrt{-\omega^2}$
$\Rightarrow p = +j\omega$, and $p = -j\omega$

The general solution is then $x = Ae^{j\omega t} + Be^{-j\omega t}$ where A and B are constants.

8. Since $e^{j\theta} = (\cos\theta + j\sin\theta)$, then:

$x = A(\cos\omega t + j\sin\omega t) + B(\cos\omega t - j\sin\omega t)$
$\Rightarrow x = (A+B)\cos\omega t + (A-B)j\sin\omega t$
$\Rightarrow x = P\cos\omega t + Q\sin\omega t$, where P and Q are arbitrary constants

1. (a) The mean position is when $h = 0.6$.

    Therefore $x = -0.4 \cos\left(\frac{\pi t}{22\,000} + \frac{\pi}{2}\right)$

    (b) Differentiating, $\dot{x} = +0.4 \times \frac{\pi}{22\,000} \sin\left(\frac{\pi t}{22\,000} + \frac{\pi}{2}\right)$

    and $\ddot{x} = 0.4 \left(\frac{\pi}{22\,000}\right)^2 \cos\left(\frac{\pi t}{22\,000} + \frac{\pi}{2}\right)$

    $\Rightarrow \ddot{x} = \left(\frac{\pi}{22\,000}\right)^2 x - x$

    $\Rightarrow \ddot{x} + \left(\frac{\pi}{22\,000}\right)^2 x = 0$

    So the motion of the deck satisfies the SHM equation with $\omega = \frac{\pi}{22\,000}$.

    (c) High tide occurs when $h = 1$,

    i.e. $\cos\left(\frac{\pi t}{22\,000} + \frac{\pi}{2}\right) = -1$

    $\Rightarrow \frac{\pi t}{22\,000} + \frac{\pi}{2} = \pi$

    $\Rightarrow t = 11\,000$ seconds $= 3$ hours 3 minutes 20 seconds

    The first high tide is at approximately 3.03 a.m.

2. $x = 0.9 \sin\left(\frac{\pi t}{3} + \frac{\pi}{12}\right)$

    (a) The time taken for one oscillation is when $\frac{\pi t}{3} = 2\pi$

    $\Rightarrow t = 6$ seconds

    The number of oscillations per second is $\frac{1}{6}$.

(continued)

(b) Differentiating twice:

$$\ddot{x} = -0.9 \times \frac{\pi^2}{9} \sin\left(\frac{\pi t}{3} + \frac{\pi}{12}\right)$$

Maximum acceleration occurs when $\sin\left(\frac{\pi t}{3} + \frac{\pi}{12}\right) = -1$

i.e. maximum acceleration $= 0.1\,\pi^2 = 0.99\ \text{ms}^{-2}$

For time:

$$\sin\left(\frac{\pi t}{3} + \frac{\pi}{12}\right) = -1$$

$$\Rightarrow \quad \frac{\pi t}{3} + \frac{\pi}{12} = \frac{3\pi}{2}$$

$$\Rightarrow \quad t = 3\left(\frac{3}{2} - \frac{1}{12}\right) = \frac{17}{4} = 4.25$$

Maximum acceleration of 0.99 ms^{-2} occurs after 4.25 seconds.

3. (a)

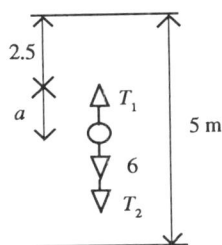

If the particle in equilibrium is a distance a metres below the centre of the two springs then:

$$6 + T_2 - T_1 = 0$$

From Hooke's law:

$$T_1 = 20\,(1.5 + a)$$

and $\quad T_2 = 20\,(1.5 - a)$

$$\Rightarrow \quad 6 + 20\,(1.5 - a) - 20\,(1.5 + a) = 0$$

$$\Rightarrow \qquad\qquad\qquad 6 - 40a = 0$$

$$\Rightarrow \qquad\qquad\qquad a = \frac{6}{40} = 0.15\ \text{metre}$$

The equilibrium position is 2.65 metres from the upper fixed point.

(b)

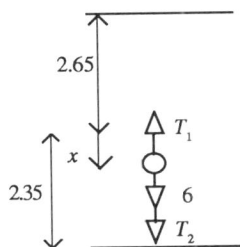

Using Newton's second law:

$$6 + T_2 - T_1 = 0.6\,\frac{d^2 x}{dt^2}$$

From Hooke's law:

$$T_1 = 20\,(1.65 + x)$$

and $\quad T_2 = 20\,(1.35 - x)$

(continued)

Then $6 + 20\,(1.35 - x) - 20\,(1.65 + x) = 0.6\,\dfrac{d^2x}{dt^2}$

$$\Rightarrow -40x = 0.6\,\frac{d^2x}{dt^2}$$

$$\Rightarrow \frac{d^2x}{dt^2} = -\frac{200}{3}x$$

This equation of motion is that for SHM,

i.e. $\dfrac{d^2x}{dt^2} = -\omega^2 x$ where $\omega = \sqrt{\left(\dfrac{200}{3}\right)}$.

So the angular frequency is $\sqrt{\left(\dfrac{200}{3}\right)}$ rad s^{-1}.

4. $\dfrac{d^2x}{dt^2} + 16\pi^2 x = 0$

The solution to the SHM equation is of the form $x = A\cos\left(\sqrt{(16\pi^2)}\,t + \varepsilon\right)$.

$$\Rightarrow x = A\cos(4\pi t + \varepsilon)$$

The initial conditions are $t = 0$, $x = 1.2$ and $\dot{x} = 0$.

So $1.2 = A\cos\varepsilon$ ①

Differentiating, $\dot{x} = -4\pi A\sin(4\pi t + \varepsilon)$

So $0 = -4\pi A\sin\varepsilon \Rightarrow \sin\varepsilon = 0 \Rightarrow \varepsilon = 0$

Substituting $\varepsilon = 0$, into ① gives $A = 1.2$

The displacement of the flag-pole is $x = 1.2\cos(4\pi t)$.

The reasons why the model may not give a realistic solution are:

• no allowance has been made for damping;

• the wind may continue to blow.

5 *Other oscillations*

5.2 Damping

> **Set up a mass oscillating on the end of a spring (a mass-spring oscillator).**
>
> (a) **Describe, by sketching a graph, how the amplitude of the oscillating mass changes with time.**
>
> (b) **What function might describe how the amplitude decreases with time?**
>
> (c) **What are the forces causing the damping?**

You may like to investigate the decrease in amplitude experimentally, by finding the amplitude at intervals of, for example, 10 seconds.

(a) The amplitude of oscillation for a spring remains reasonably constant, but over time the amplitude gradually decreases and at some point the oscillations will cease.

Possible graphs of amplitude against time are:

(i) 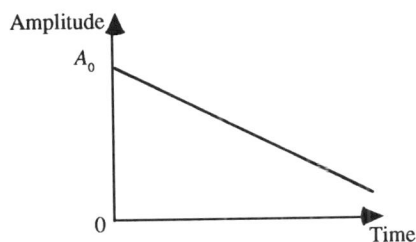   (ii)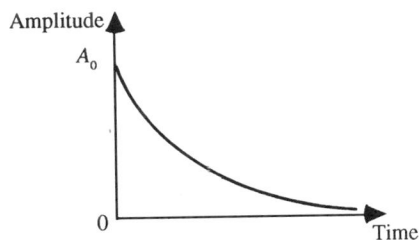

(b) Possible functions could be: (i) amplitude $= A_0 - \lambda t$
(ii) amplitude $= A_0 \, e^{-\lambda t}$

where A_0 is the initial amplitude and λ is a constant.

From experiment it is found that $A_0 \, e^{-\lambda t}$ is the more appropriate function for modelling the amplitude decay.

(c) Possible forces are:

- the air resistance due to the mass moving through air;
- the frictional force at the support;
- internal forces in the spring itself.

Pendulum clocks

For the best experimental results:

- the string of the pendulum should be fixed at one end so that it swings freely;
- the string should be long enough for the size of the bob to be insignificant (i.e. it can be considered to be a particle);
- use a fairly light string or thread so that its weight is negligible compared with the bob;
- time a number of oscillations, for example 5 or 10, and calculate an average time period (for small amplitude oscillations the amplitude is reasonably constant for a few oscillations, though for larger initial angles of displacement the amplitude decreases quickly);
- take care that when investigating the effects of one variable the others are kept constant.

Possible observations are:

Mass: Provided the mass is sufficiently large so that the string can be assumed to be light, you should find that increasing the mass has little or no effect on the amplitude. This can be tested by comparing the time period for a light bob with that for a heavy bob. The time period appears to be independent of mass.

Length: The time period increases as the length of the pendulum increases. The graph shows time period against length for possible data.

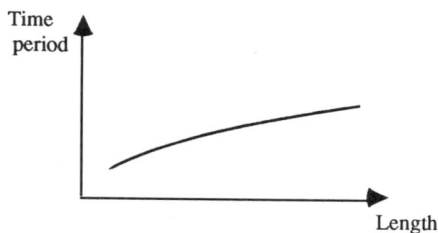

From the graph, a possible relationship is that : $\tau \propto \sqrt{\text{length}}$. You should be able to use your data to find the constant of proportionality. Note that for small lengths it is difficult to collect reliable data as the size of the mass or bob is no longer insignificant.

Amplitude: For small initial angles of displacement the oscillations are similar and the amplitude does not decrease noticeably. Provided the initial angle is small (it does not exceed, for example, $\frac{\pi}{12}$ radians or 15°) then the time period appears to be unaffected.

For larger initial angles, the time period appears to increase as the initial angle increases. This is more noticeable for angles greater than $\frac{\pi}{4}$ radians. However, as was stated earlier, it is more difficult to obtain reliable results for large amplitude oscillations as the amplitude decreases quite quickly.

Calculating acceleration

1. $\overrightarrow{OP} = \begin{bmatrix} r\cos\theta \\ r\sin\theta \end{bmatrix} = r\begin{bmatrix} \cos\theta \\ \sin\theta \end{bmatrix}$

   $\begin{bmatrix} \cos\theta \\ \sin\theta \end{bmatrix}$ is therefore in the direction $\overrightarrow{OP}$. Its length is $\sqrt{(\sin^2\theta + \cos^2\theta)} = 1$.

2. $\mathbf{r} = r\begin{bmatrix} \cos\theta \\ \sin\theta \end{bmatrix}$

   $\mathbf{v} = \dfrac{d\mathbf{r}}{dt} = r\begin{bmatrix} -\sin\theta \times \dot\theta \\ \cos\theta \times \dot\theta \end{bmatrix} = r\dot\theta\begin{bmatrix} -\sin\theta \\ \cos\theta \end{bmatrix}$

   $\begin{bmatrix} -\sin\theta \\ \cos\theta \end{bmatrix}$ is a unit vector and so $\mathbf{v}$ has magnitude $r\dot\theta$.

3. (a) Acceleration, $\mathbf{a} = \dfrac{d\mathbf{v}}{dt} = r\begin{bmatrix} -\cos\theta \times \dot\theta^2 - \sin\theta \times \ddot\theta \\ -\sin\theta \times \dot\theta^2 + \cos\theta \times \ddot\theta \end{bmatrix}$

   $\Rightarrow \mathbf{a} = -r\dot\theta^2\begin{bmatrix} \cos\theta \\ \sin\theta \end{bmatrix} + r\ddot\theta\begin{bmatrix} -\sin\theta \\ \cos\theta \end{bmatrix}$

   (b) $\begin{bmatrix} -\sin\theta \\ \cos\theta \end{bmatrix}$ has length $\sqrt{(\sin^2\theta + \cos^2\theta)} = 1$

   $\begin{bmatrix} \cos\theta \\ \sin\theta \end{bmatrix} \cdot \begin{bmatrix} -\sin\theta \\ \cos\theta \end{bmatrix} = -\cos\theta\sin\theta + \cos\theta\sin\theta = 0$

   Therefore $\begin{bmatrix} -\sin\theta \\ \cos\theta \end{bmatrix}$ is a unit vector in a direction perpendicular to $\overrightarrow{OP}$.

Damped SHM

1. $\dot{x} = Ape^{pt}$ and $\ddot{x} = Ap^2e^{pt}$

     $\Rightarrow mAp^2e^{pt} + CApe^{pt} + kAe^{pt} = 0$

     $\Rightarrow mp^2 + Cp + k = 0$

2. $mp^2 + 20p + 10 = 0$

     $\Rightarrow p = \dfrac{-20 \pm \sqrt{(400 - 40m)}}{2m}$

The auxiliary equation has real, equal or imaginary roots depending on the value of $\sqrt{(400 - 40m)}$. If:

     $400 - 40m > 0$ (i.e. $m < 10$) the roots are real;

     $400 - 40m = 0$ (i.e. $m = 10$) the roots are equal;

     $400 - 40m < 0$ (i.e. $m > 10$) the roots are imaginary.

3. $7.5\ddot{x} + 20\dot{x} + 10x = 0$

     $\Rightarrow 7.5p^2 + 20p + 10x = 0$

     $\Rightarrow p = -\dfrac{2}{3}$ or -2

     $\Rightarrow x = Ae^{-\frac{2}{3}t} + Be^{-2t}$

When $t = 0$, $x = 2$ and $\dot{x} = 0$

So $2 = A + B$ and $0 = -\dfrac{2}{3}A - 2B$

Then $A = 3$ and $B = -1$:

     $x = 3e^{-\frac{2}{3}t} - e^{-2t}$

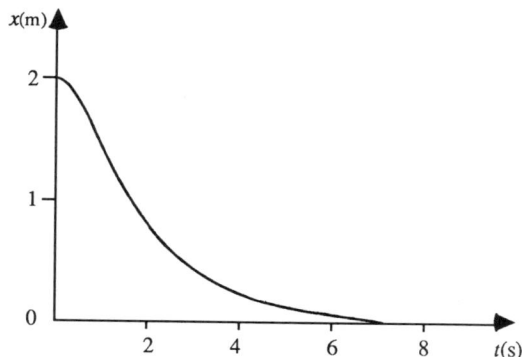

The mass has an initial displacement of 2 metres. When released it slowly returns to its equilibrium position. It does not oscillate and takes approximately 7 seconds to return to equilibrium.

(continued)

4.
$$10\ddot{x} + 20\dot{x} + 10x = 0$$
$$\Rightarrow \quad 10p^2 + 20p + 10 = 0$$
$$\Rightarrow \quad p = -1 \text{ (repeated)}$$
$$\Rightarrow \quad x = (A + Bt)\,e^{-t}$$

When $t = 0$, $x = 2$ and $\dot{x} = 0$

So $2 = A$ and $-A + B = 0$

Then $x = (2 + 2t)\,e^{-t}$

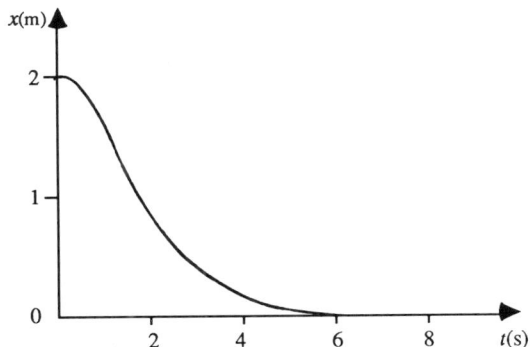

As with the overdamped case, the mass does not oscillate but returns to equilibrium. In this case, the displacement initially decreases more slowly but, after 2.8 seconds, the displacement decreases more quickly and equilibrium is reached faster.

5.
$$100\ddot{x} + 20\dot{x} + 10x = 0$$
$$\Rightarrow \quad 100p^2 + 20p + 10 = 0$$
$$\Rightarrow \quad p = -0.1 \pm 0.3j$$
$$\Rightarrow \quad x = Ae^{-0.1t}\cos(0.3t + \varepsilon)$$

When $t = 0$, $x = 2$ and $\dot{x} = 0$

So $2 = A\cos\varepsilon$ and $0 = -0.1A\cos\varepsilon - 0.3A\sin\varepsilon$

Then $x = 2.11\,e^{-0.1t}\cos(0.3t - 0.322)$

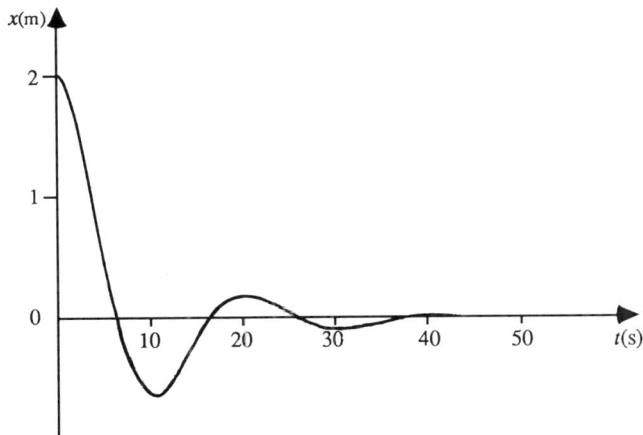

The mass oscillates about the equilibrium position, performing approximately two complete oscillations. The time period is about 20 seconds. The amplitude decreases quickly.

6.
As $C \to 0$, $n \to \sqrt{\left(\dfrac{k}{m}\right)}$ and $x \to A\cos\left(\sqrt{\left(\dfrac{k}{m}\right)}t + \varepsilon\right)$

This is the solution for SHM of a mass-spring oscillator.

1.   (a) $\tau = 2\pi \sqrt{(\frac{l}{g})}$

   For the Earth: $1 = 2\pi \sqrt{(\frac{l}{9.81})} \Rightarrow l = 0.248$

   On the Moon: $\tau = 2\pi \sqrt{(\frac{0.248}{1.62})} = 2.46$

   The time period on the surface of the Moon is 2.5 seconds.

   $\left(\text{Alternatively, } \dfrac{\tau_{\text{Earth}}}{\tau_{\text{Moon}}} = \sqrt{(\frac{1.62}{9.81})} \Rightarrow \tau_{\text{Moon}} = 2.46\right)$

   (b)   The model assumes that air resistance is negligible. This is a more valid assumption on the Moon than on Earth.

2. $x = ae^{-t} \cos(\omega t)$

   (a)   Initially $t = 0$ and $x = 0.01 \Rightarrow a = 0.01$

   $$\tau = 0.01 = \frac{2\pi}{\omega} \quad \Rightarrow \omega = 200\pi$$

   (b) $\dot{x} = -a\omega e^{-t} \sin \omega t - ae^{-t} \cos \omega t$ ①

   $\Rightarrow \ddot{x} = 2a\omega e^{-t} \sin \omega t - a\omega^2 e^{-t} \cos \omega t + ae^{-t} \cos \omega t$

   $\Rightarrow \ddot{x} = 2a\omega e^{-t} \sin \omega t + (1 - \omega^2) ae^{-t} \cos \omega t$ ②

   $\Rightarrow \ddot{x} + 2\dot{x} + (\omega^2 + 1) x = 0$ 2① + ②

   Hence $x = ae^{-t} \cos \omega t$ is a solution to the damped SHM equation.

3.   (a) $x = \dfrac{at}{2\omega} \sin \omega t$

   $\Rightarrow \dot{x} = \dfrac{a}{2\omega} \sin \omega t + \dfrac{at}{2} \cos \omega t$

   $\Rightarrow \ddot{x} = \dfrac{a}{2} \cos \omega t + \dfrac{a}{2} \cos \omega t - \dfrac{a\omega t}{2} \sin \omega t$

   $\Rightarrow \ddot{x} = a \cos \omega t - \omega^2 \left(\dfrac{at}{2\omega} \sin \omega t \right)$

   $\Rightarrow \ddot{x} = a \cos \omega t - \omega^2 x$

   $\Rightarrow \ddot{x} + \omega^2 x = a \cos \omega t.$

   Thus $x = \dfrac{at}{2\omega} \sin \omega t$ is a particular integral.

(continued)

(b)

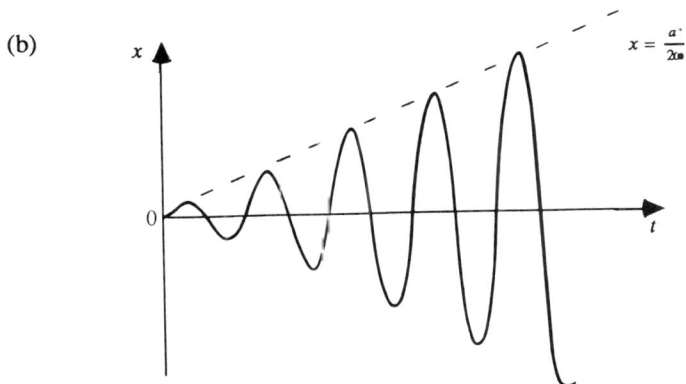

$$x = \frac{a}{2\omega}$$

This is the solution when resonance occurs.

The amplitude is growing linearly with time, where $a(t) = \frac{a\,t}{2\omega}$.

This would be invalid for large time. Unless there were damping forces to control the amplitude growth, the harp would shatter.

4. (a) The algebra involved in obtaining an analytical solution is complicated unless the small amplitude approximation, $\sin \theta \approx \theta$, is made.

    (b) A numerical solution is prone to errors. The solution is not general but only applies to particular initial conditions.